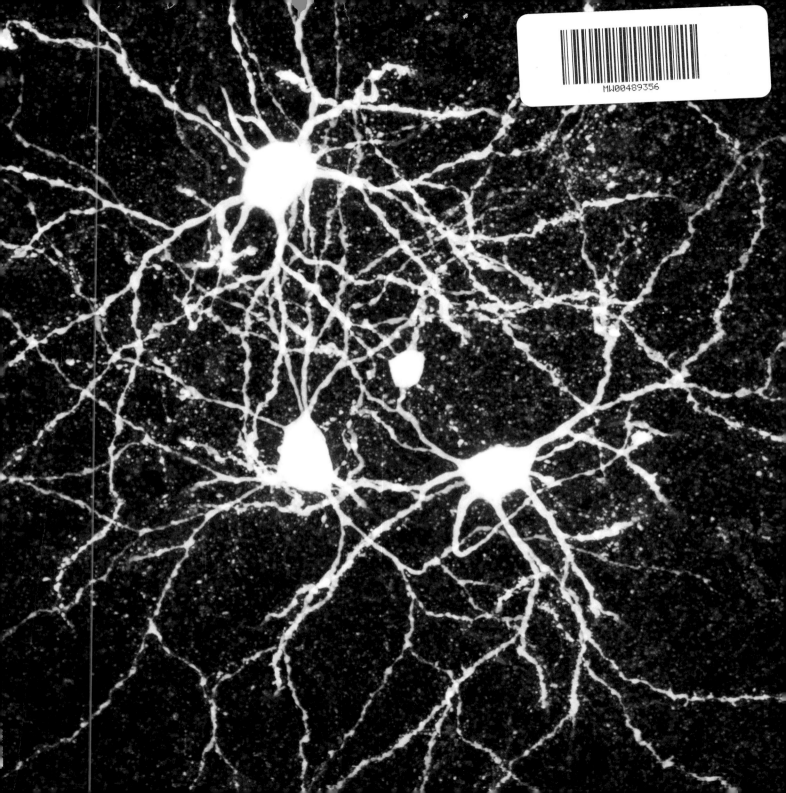

NEURON GALAXY

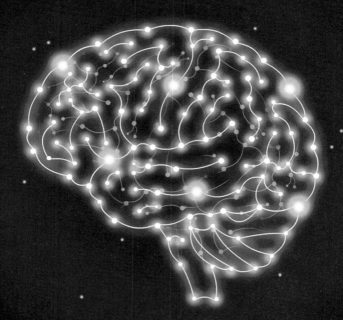

A Story from Morphonix about Your Brain

Neuron Galaxy is part of *NeuroPlay Adventures*, a suite of games, apps, stories, and songs that help children understand their growing brains.

NeuroPlay Adventures Advisors

Floyd Bloom, MD, Professor Emeritus, The Scripps Research Institute

Adam Gazzaley, MD, PhD, Professor of Neurology, Physiology and Psychiatry
Director, Neuroscience Imaging Center, University of California, San Francisco

• • •

*We would like to acknowledge the additional help and advice of the
following neuroscientists in preparing this book. Any errors are ours alone.*

Eric H. Chudler, Research Associate Professor, University of Washington, Seattle, WA

Loren Frank, PhD, Professor, Howard Hughes Medical Institute
and University of California, San Francisco

Mark N. Miller, PhD, Brainard Lab, Howard Hughes Medical Institute
and University of California, San Francisco

NEURON GALAXY

A Story from Morphonix about Your Brain

A NeuroPlay Adventure Story from Morphonix

Karen Littman, Creative Director

Written by Jay Leibold

Illustrated by Max Weinberg and Christine Gralapp

Published by Morphonix LLC

Image of neurons in front of book used by permission of Mark N. Miller, Brainard Lab, Howard Hughes Medical Institute and University of California, San Francisco.

Image of simulated evolution of the universe in back of book courtesy of V. Springel and the Virgo Consortium, Max Planck Institute for Astrophysics.

• • •

www.morphonix.com

ISBN-13: 978-0692747667
ISBN-10: 0692747664

Library of Congress Control Number: 2016912057
Morphonix, LLC, San Rafael, CA

Morphonix is an award-winning developer of learning apps, games, songs, and stories that teach children about neuroscience and their growing brains in a fun and engaging way. Awards for previous games include Common Sense Media ON for Learning Award and Parents' Choice Gold Award.

Development of this project was supported by the National Institute of Mental Health of the National Institutes of Health under Award #R44MH096339. The content is solely the responsibility of the developers and does not represent the official views of The National Institutes of Health.

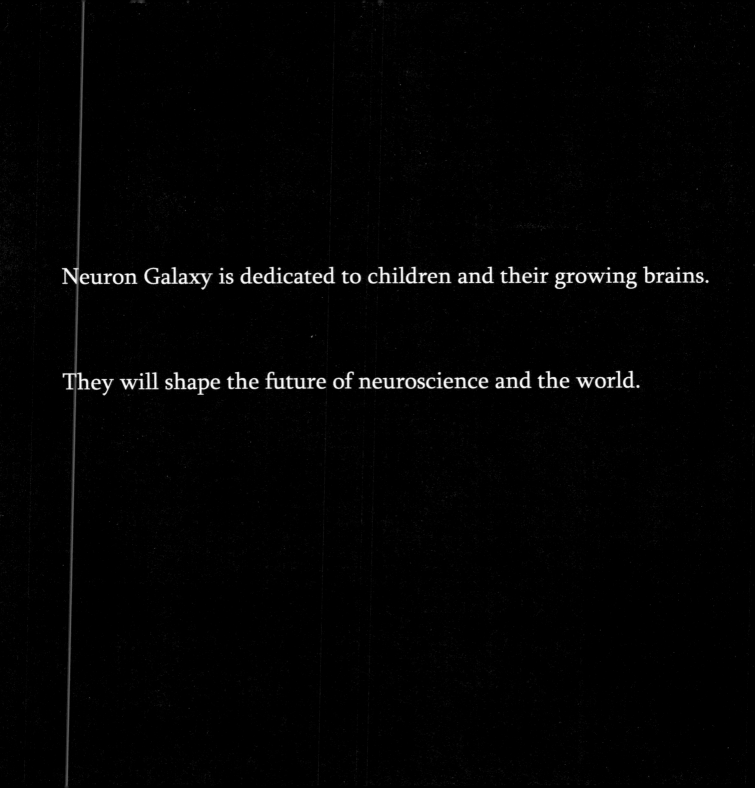

Neuron Galaxy is dedicated to children and their growing brains.

They will shape the future of neuroscience and the world.

When you were very, very little,

a tiny baby neuron grew inside your head.

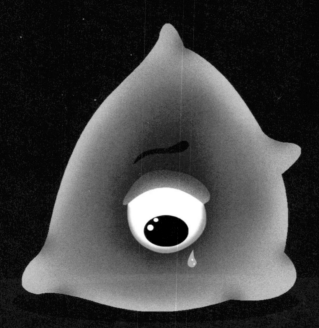

The baby neuron was lonely.

It wanted to connect to neuron friends.

The baby neuron sent out branches.

The branches went and felt around for other neurons.

Look! The branches found a neuron friend!

All they had to do was reach out.

What joy!

The neurons reached out to find more friends.

And they found them.

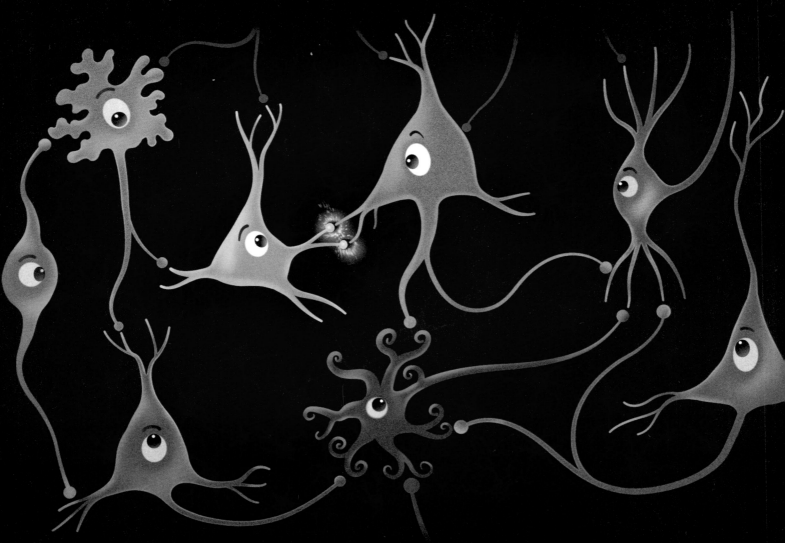

More and more of them!

All the neurons talked and shared information.

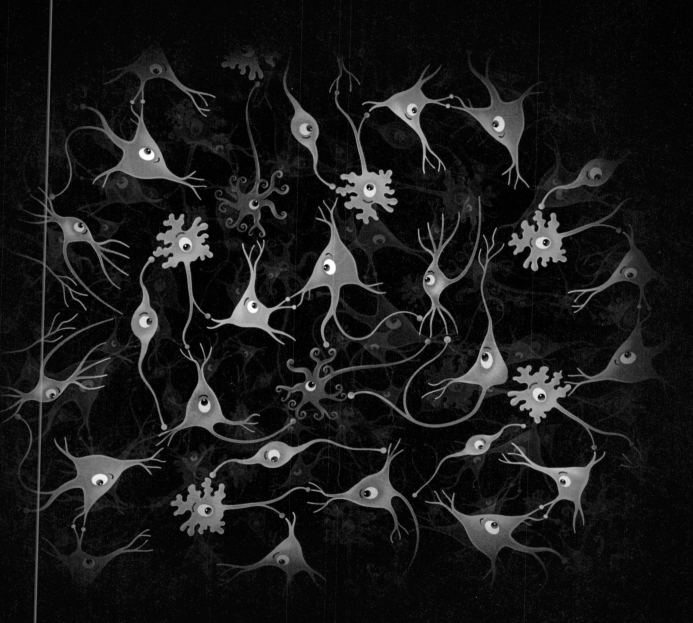

The more they talked, the more new friends they found!

They formed many pathways in a big network.

Lots and lots and lots of different kinds of neurons connected.

Each type had a different job, like members of a team.

But what were they all talking about?

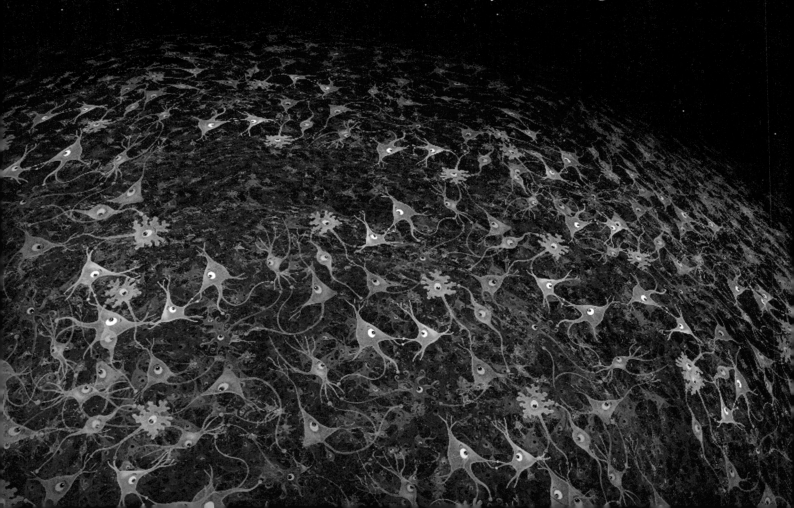

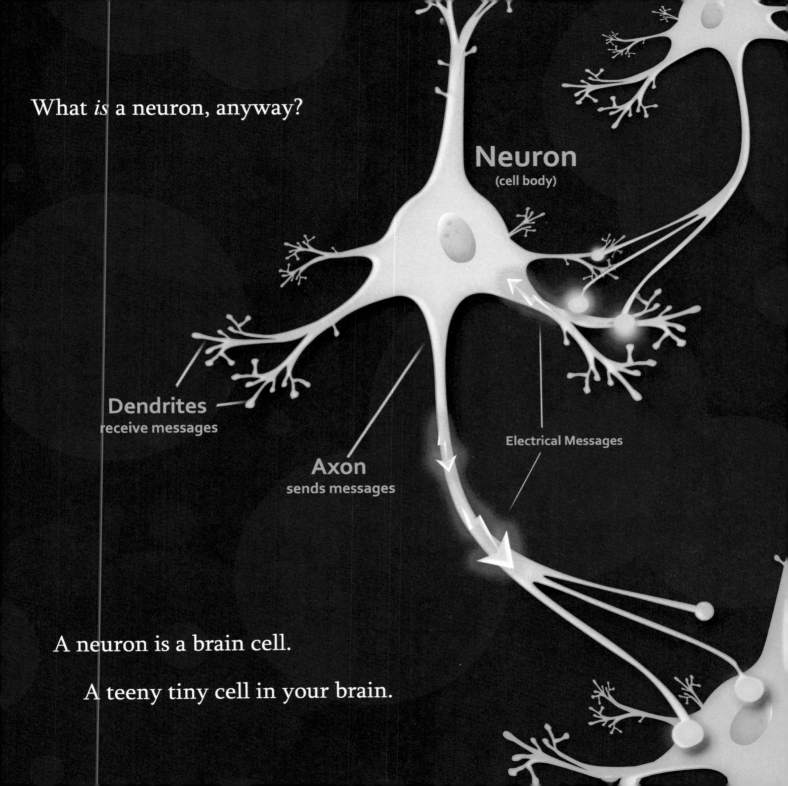

What *is* a neuron, anyway?

Neuron
(cell body)

Dendrites
receive messages

Axon
sends messages

Electrical Messages

A neuron is a brain cell.

A teeny tiny cell in your brain.

How teeny tiny is a neuron?

So tiny you can't see it with your eye.

You need a powerful microscope.

So tiny you could put 500 of them in a space as small

as the dot on this letter i

400x

2000x

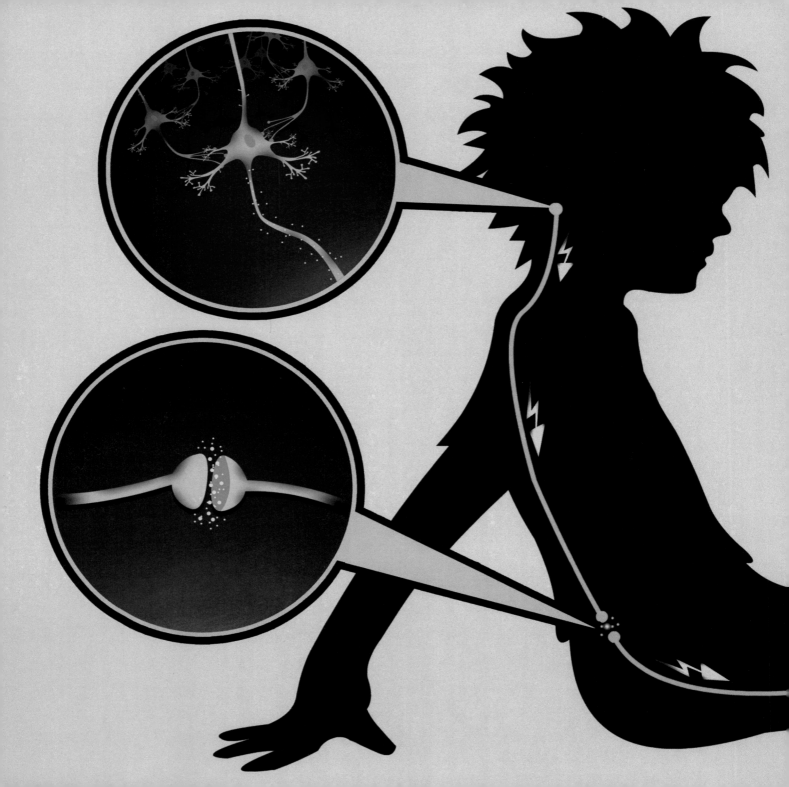

The cell body of a neuron might be small,

but its main branch, which is called an axon, can be long.

As long as three feet!

That's so it can reach other neurons.

And make your toes wiggle!

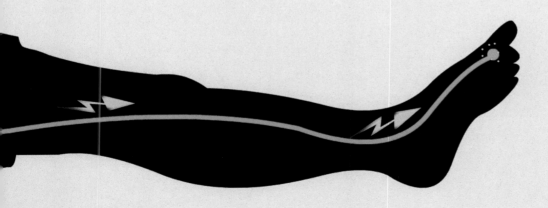

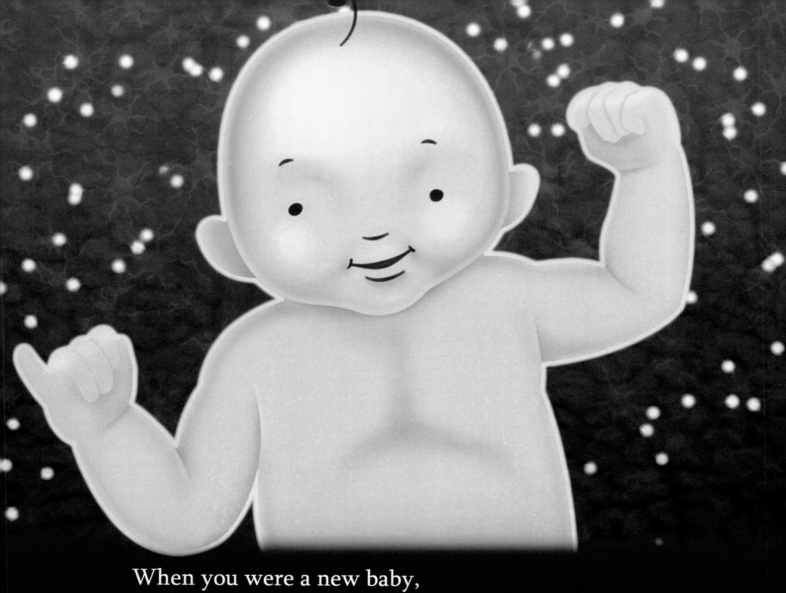

When you were a new baby,

you could only say *goo goo goo* and *ga ga ga*.

You could wave your arms. You could wiggle around.

You waited for Mom and Dad to feed you.

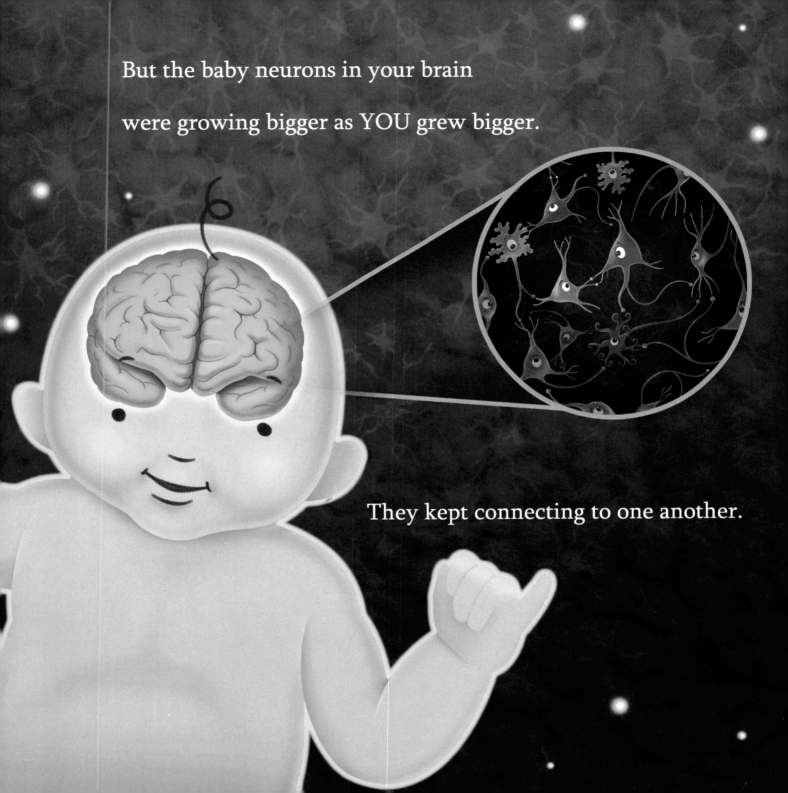

But the baby neurons in your brain

were growing bigger as YOU grew bigger.

They kept connecting to one another.

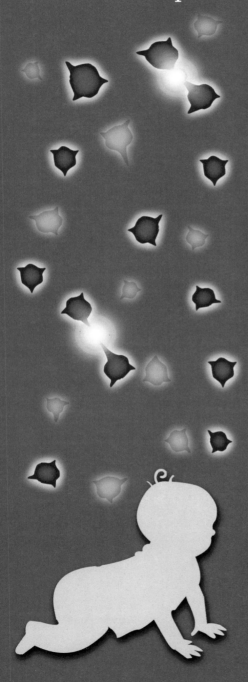

Pretty soon you
could stand up . . .

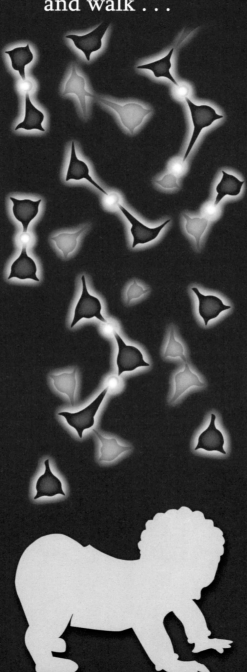

and walk . . .

and say hello!

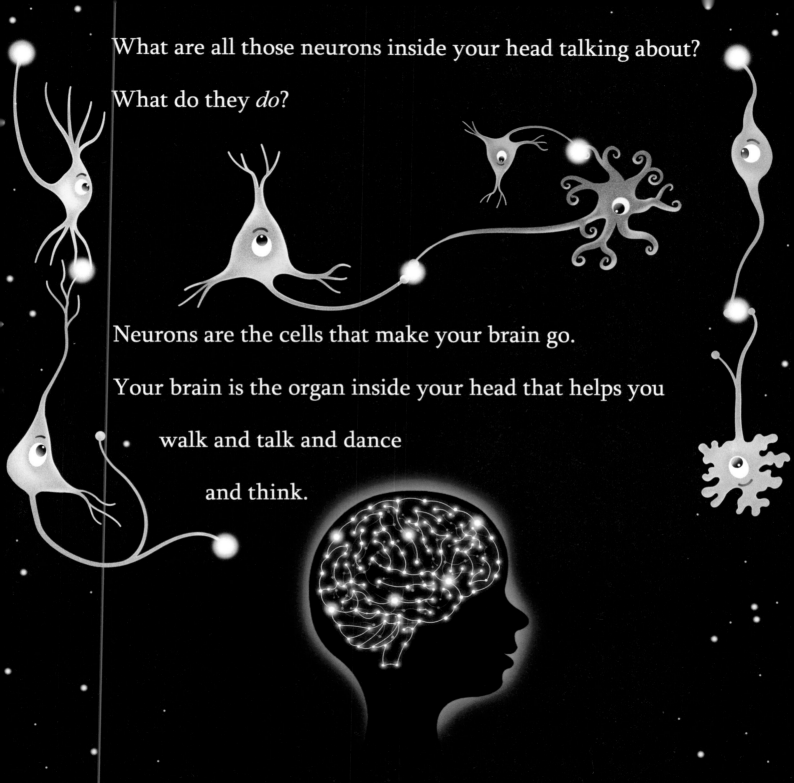

What are all those neurons inside your head talking about?

What do they *do*?

Neurons are the cells that make your brain go.

Your brain is the organ inside your head that helps you

walk and talk and dance

and think.

Because of your neurons,

because their branches connect like shaking hands . . .

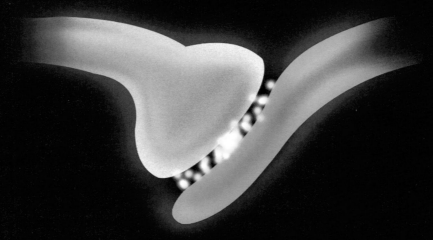

You can see your friend . . .

and say hello . . .

and shake hands.

Because of your neurons, you can learn new things in school.

Because of your neurons, you can remember what you did last week.

Because of your neurons, you can draw a story.

Because of your neurons, you can ride a bike.

Because of your neurons, you can breathe and swallow.

Because of your neurons, you can feel happy or sad.

Every time you do just ONE of these things,

tens of millions of neurons spark and connect in your brain.

How many neurons are in your brain?

LOTS. 100 billion. That's 100 thousand

thousand

thousand.

That's as many stars as are in our Milky Way galaxy!

By the time you grow up, there will be so many axons connecting your neurons that if you put them end to end, they would circle the Earth . . .

4½ times!!!

As you grow up, the neurons in your brain

keep shifting and shaping their connections.

Some pathways grow stronger and some go away.

It all depends on how you use your brain.

When you read books or learn a new dance,

those neuron pathways grow stronger.

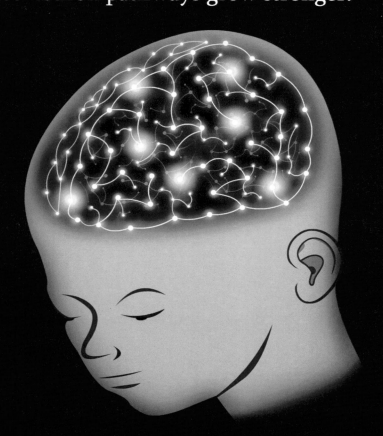

The unique and special ways your neurons connect

are what make you YOU.

There's no one else exactly like you in the world.

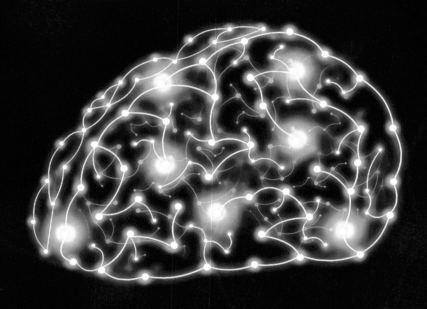

You can thank your brain for that.

And don't forget to take care of all those hard-working neurons

and keep them healthy, because they take care of you.

As you go through school . . .

and get to know

your neighborhood . . .

And travel to other places . . .

You shape your brain and your brain shapes you.

Your brain is one of the most amazing and
wonderful things on Earth.
It helps you understand how
amazing and beautiful our Earth is . . .

And our solar system . . .

And our galaxy.

Remember, you have as many neurons as there are stars in our galaxy.

If you take care of them and help them make new connections,

they will help you appreciate your friends and the awe and wonder

of the planet, the galaxy, and the vast universe.

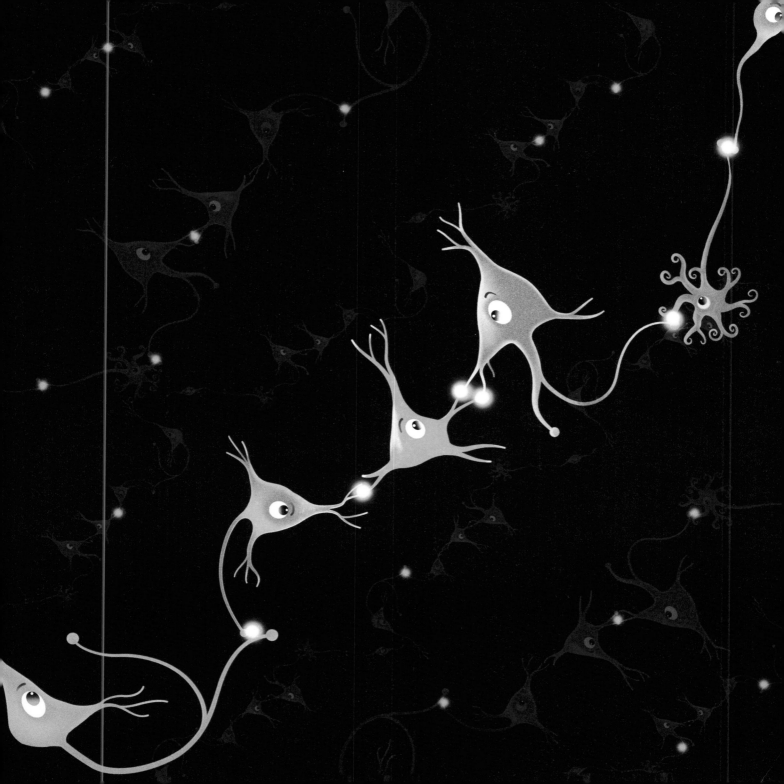

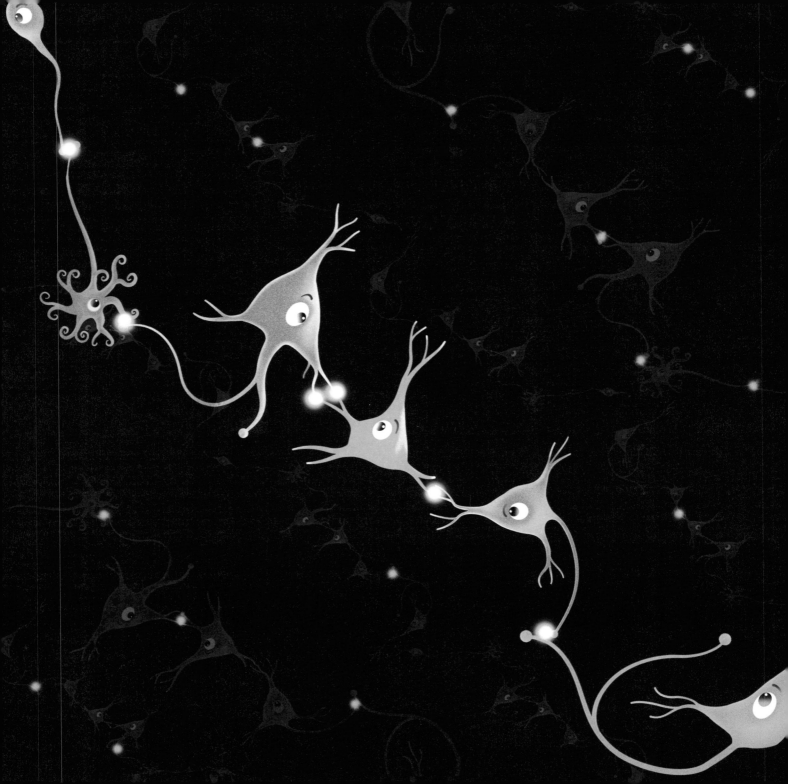

Neuron Galaxy is part of *NeuroPlay Adventures*,
a suite of games, apps, stories, and songs
that help children understand their growing brains.

Apps from Morphonix:

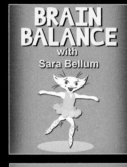   

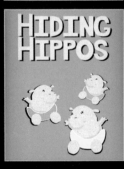   

Learn more at
www.morphonix.com

Download the free interactive version of Neuron Galaxy here:
www.morphonix.com/products/neuron-galaxy

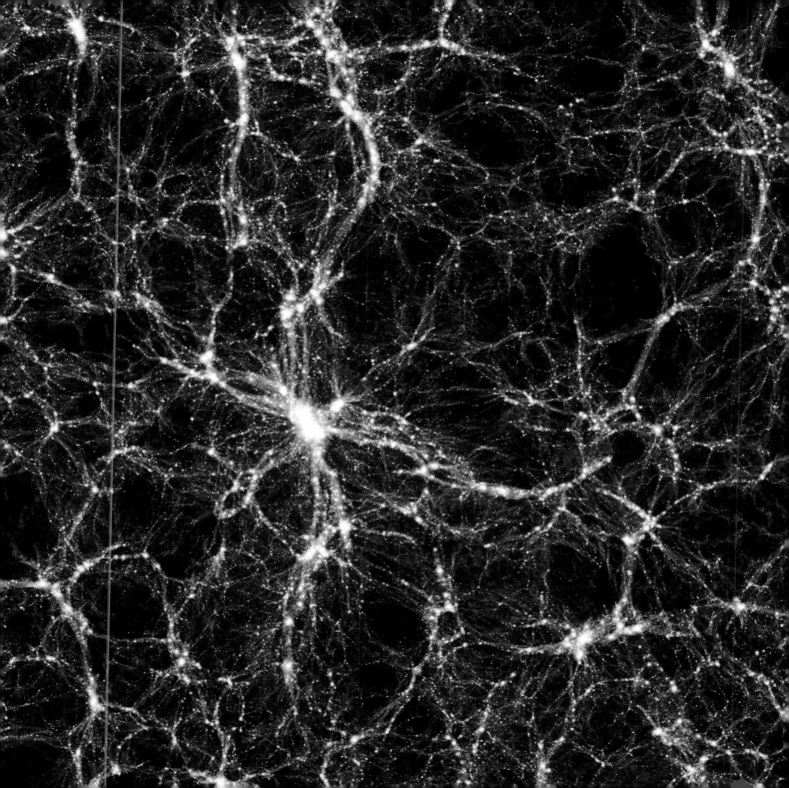

Made in the USA
San Bernardino, CA
25 January 2017